300

EXCITING FACTS

ABOUT SUPERCARS

FOR CURIOUS KIDS

Dive into the pulse-pounding universe of high-performance vehicles with '300 Exciting Facts about Supercars.' This captivating book takes you on a thrilling ride through the rich histories, record-breaking feats, and cutting-edge technology that define the world of supercars. Whether you're a seasoned enthusiast or a newcomer, this book is your key to unlocking the fascinating stories and unparalleled engineering behind these extraordinary machines. Buckle up for an exhilarating journey through the heart and soul of automotive excellence!"

LIST OF SUPERCARS

BUGATTI VEYRON
LAMBORGHINI AVENTADOR
FERRARI LAFERRARI
MCLAREN P1
PORSCHE 918 SPYDER
KOENIGSEGG JESKO
ASTON MARTIN VALKYRIE
PAGANI HUAYRA
RIMAC C_TWO
AUDI R8
FORD GT
LEXUS LFA
MERCEDES-AMG ONE
LOTUS EVIJA
CHEVROLET CORVETTE ZR1
JAGUAR C-X75
MCLAREN SPEEDTAIL
FERRARI SF90 STRADALE
LAMBORGHINI HURACAN PERFORMANTE
PORSCHE 911 GT2 RS
BUGATTI CHIRON
KOENIGSEGG GEMERA
PAGANI ZONDA
ASTON MARTIN ONE-77
RIMAC CONCEPT_ONE

MCLAREN 765LT
LOTUS EXIGE CUP 430
FORD MUSTANG SHELBY GT500
LAMBORGHINI CENTENARIO
FERRARI 488 PISTA
PORSCHE CARRERA GT
MCLAREN 600LT
AUDI RS E-TRON GT
PAGANI IMOLA
KOENIGSEGG REGERA
BUGATTI DIVO
ASTON MARTIN DBS SUPERLEGGERA
LAMBORGHINI SIAN
FERRARI F8 TRIBUTO
PORSCHE 911 GT3 RS
MCLAREN GT
RIMAC NEVERA
LOTUS ELISE CUP 260
FORD GT MK II
LAMBORGHINI VENENO
BUGATTI CENTODIECI
KOENIGSEGG AGERA RS
PAGANI HUAYRA ROADSTER
ASTON MARTIN VULCAN
MCLAREN SABRE

BUGATTI VEYRON

- It held the title of the world's fastest car.

- The Super Sport version set a top speed record of over 267 mph.

- Each Veyron had a distinctive "EB" logo on the back.

- It featured a massive 8.0-liter W16 engine.

- The Veyron's tires had to withstand extreme speeds, leading to their high cost.

LAMBORGHINI AVENTADOR

- **Named after a legendary fighting bull.**

- **Its V12 engine produces a powerful roar.**

- **The Aventador can accelerate from 0 to 60 mph in under 3 seconds.**

- **It features scissor doors, a Lamborghini trademark.**

- **Aventador's carbon fiber monocoque provides strength and lightness.**

FERRARI LAFERRARI

- **LaFerrari means "The Ferrari" in Italian.**

- **It's a hybrid hypercar with a V12 engine and electric motor.**

- **The car's body design enhances aerodynamics for speed.**

- **LaFerrari has a unique active aerodynamic system.**

- **Only 499 units were produced, making it highly exclusive.**

MCLAREN P1

- Part of the "holy trinity" of hybrid hypercars.

- The P1 uses an electric motor for instant torque.

- Features an adjustable rear wing for aerodynamics.

- Accelerates from 0 to 60 mph in less than 2.8 seconds.

- McLaren P1's top speed is electronically limited to 217 mph.

PORSCHE 918 SPYDER

- It's a plug-in hybrid supercar.

- The 918 Spyder set a Nürburgring lap record for production cars.

- Uses a combination of a V8 engine and electric motors.

- The "Weissach Package" reduces weight and enhances performance.

- Limited production, with only 918 units built.

KOENIGSEGG JESKO

- **Named after Jesko von Koenigsegg, the father of the company's founder.**

- **Features the innovative Koenigsegg Light Speed Transmission (LST).**

- **The Jesko Absolut variant aims for top speed records.**

- **It has a distinctive aerodynamic design, including a large rear wing.**

- **The engine produces over 1,600 horsepower.**

ASTON MARTIN VALKYRIE

- **Developed in collaboration with Red Bull Racing.**

- **Designed by F1 aerodynamics expert Adrian Newey.**

- **Limited to just 150 units, ensuring exclusivity.**

- **The V12 engine is supplied by Cosworth.**

- **Features a unique interior with a center-mounted driver's seat.**

PAGANI HUAYRA

- **Named after the Incan god of wind, Huayra Tata.**

- **The Huayra uses a Mercedes-AMG V12 engine.**

- **It incorporates advanced aerodynamics, including active flaps.**

- **Pagani is known for bespoke customization, making each Huayra unique.**

- **Limited production, with around 100 units produced.**

RIMAC C_TWO

- **An all-electric hypercar with a focus on performance.**

- **Accelerates from 0 to 60 mph in under 1.9 seconds.**

- **The C_Two has a top speed exceeding 250 mph.**

- **Rimac specializes in electric vehicle technology.**

- **Features advanced driver-assistance systems.**

AUDI R8

- The R8 is Audi's first supercar.

-

- Available in both V8 and V10 engine configurations.

-

- Audi R8's mid-engine design contributes to balanced handling.

-

- The R8 has won several prestigious awards for its design.

-

- It shares its platform with the Lamborghini Huracan.

FORD GT

- **The modern GT pays homage to the iconic Ford GT40.**

- **Features a lightweight carbon fiber body for enhanced performance.**

- **The GT's engine is a twin-turbocharged V6 EcoBoost.**
- **It won its class at the 2016 24 Hours of Le Mans.**

- **Limited production, with a strict application process for buyers.**

LEXUS LFA

- **The LFA is Lexus's first and only supercar.**

- **It boasts a high-revving V10 engine with a unique exhaust note.**

- **Carbon fiber reinforced polymer (CFRP) construction enhances rigidity.**

- **Lexus limited production to maintain exclusivity, building only 500 units.**

- **The LFA underwent extensive testing at the Nürburgring.**

MERCEDES-AMG ONE

- **The "Project One" hypercar brings Formula 1 technology to the road.**

- **Powered by a turbocharged V6 engine paired with electric motors.**

- **The design is inspired by Mercedes-AMG's F1 cars.**

- **Limited to just 275 units, ensuring rarity.**

- **Offers a top speed exceeding 217 mph.**

LOTUS EVIJA

- **The Evija is Lotus's first all-electric hypercar.**

- **It boasts a striking aerodynamic design with Venturi tunnels.**

- **The Evija's electric powertrain produces over 1,900 horsepower.**

- **Features a unique charging system for quick battery replenishment.**

- **Limited to a production run of 130 units.**

CHEVROLET CORVETTE ZR1

- **The ZR1 is the most powerful production Corvette.**

- **It features a supercharged V8 engine with over 750 horsepower.**

- **The ZR1 set a lap record at Virginia International Raceway.**

- **Known for its distinctive and aggressive styling.**

- **Offers a high-performance track package.**

JAGUAR C-X75

- **Originally a concept, the C-X75 was featured in the James Bond film "Spectre."**

- **It was intended to be a hybrid supercar with a gas-turbine range extender.**

- **Jaguar planned to produce a limited run, but it remained a concept car.**

- **The C-X75's design drew inspiration from Jaguar's racing heritage.**

- **Despite not entering production, it showcased advanced technology.**

MCLAREN SPEEDTAIL

- **Known as the "Hyper-GT," the Speedtail is McLaren's fastest car.**

- **It features a central driving position, similar to the McLaren F1.**

- **The Speedtail's top speed exceeds 250 mph.**

- **Limited production of 106 units, all of which were pre-sold.**

- **It incorporates innovative aerodynamics for optimal performance.**

FERRARI SF90 STRADALE

- **The SF90 Stradale is Ferrari's first plug-in hybrid production car.**

- **Combines a V8 engine with three electric motors for a total of 1,000 horsepower.**

- **Features advanced aerodynamics, including a redesigned front end.**

- **It can accelerate from 0 to 60 mph in under 2.5 seconds.**

- **The SF90 Stradale is positioned as a flagship model for Ferrari.**

Lamborghini Huracan Performante

- **The Performante is known for its active aerodynamics system called ALA.**

- **ALA enhances downforce and aerodynamic efficiency during different driving conditions.**

- **It set a production car lap record at the Nürburgring Nordschleife.**

- **The Performante's V10 engine produces over 600 horsepower.**

- **Features lightweight materials for improved performance.**

PORSCHE 911 GT2 RS

- **The GT2 RS is the most powerful 911 model ever built.**

- **It features a twin-turbocharged flat-six engine with over 700 horsepower.**

- **The GT2 RS holds the Nürburgring lap record for production cars.**

- **Known for its rear-wheel-drive layout and precise handling.**

- **Limited production with a focus on track performance.**

Bugatti Chiron

- **Successor to the Veyron, the Chiron is known for its extreme performance.**

- **It features a quad-turbocharged W16 engine delivering over 1,500 horsepower.**

- **The Chiron's top speed is electronically limited to 261 mph.**

- **Bugatti offers extensive customization options for Chiron buyers.**

- **Limited production, with only 500 units planned.**

KOENIGSEGG GEMERA

- **The Gemera is Koenigsegg's first four-seater hypercar.**

- **It features a unique 3-cylinder Freevalve engine combined with electric motors.**

- **The Gemera emphasizes both performance and practicality.**

- **Koenigsegg's "Tiny Friendly Giant" (TFG) engine is central to its design.**

- **Limited production, with only 300 units planned.**

PAGANI ZONDA

- **The Zonda was Pagani's first production car.**

- **Known for its bespoke craftsmanship and attention to detail.**

- **Each Zonda is custom-built to the buyer's specifications.**

- **Despite production officially ending, new one-off models occasionally emerge.**

- **The Zonda played a significant role in establishing Pagani as a hypercar manufacturer.**

ASTON MARTIN ONE-77

- **Named for its limited production run of 77 units.**

- **The One-77 features a naturally aspirated V12 engine.**

- **Aston Martin collaborated with various luxury brands for bespoke interior options.**

- **It has a distinctive aluminum body and a lightweight carbon fiber monocoque.**

- **The One-77 boasts a top speed of over 220 mph.**

RIMAC CONCEPT_ONE

- **The Concept_One is an all-electric hypercar from Rimac.**

- **It gained fame after being featured in "The Grand Tour" TV series.**

- **Accelerates from 0 to 60 mph in just 2.5 seconds.**

- **Rimac specializes in electric vehicle technology and components.**

- **Limited production, with only a handful of units produced.**

MCLAREN 765LT

- **The 765LT is a track-focused version of the McLaren 720S.**

- **LT stands for "Longtail," a homage to McLaren's racing history.**

- **It features a twin-turbocharged V8 engine with over 750 horsepower.**

- **The 765LT has aerodynamic enhancements for increased downforce.**

- **Limited production, with a focus on lightweight performance.**

Lotus Exige Cup 430

- **The Exige Cup 430 is one of Lotus's high-performance models.**

- **It features a supercharged V6 engine with 430 horsepower.**

- **Known for its lightweight design and exceptional handling.**

- **The Cup 430 has aggressive aerodynamics, including a large rear wing.**

- **Lotus emphasizes the Exige's connection to its motorsport heritage.**

Ford Mustang Shelby GT500

- **The GT500 is the most powerful street-legal Ford Mustang.**

- **It features a supercharged V8 engine with over 700 horsepower.**

- **The GT500 has a dual-clutch transmission for rapid gear changes.**

- **It incorporates advanced aerodynamics for improved performance.**

- **Known for its iconic Shelby Cobra emblem and racing heritage.**

Lamborghini Centenario

- Built to celebrate the 100th birthday of Lamborghini's founder.

- Limited production with only 40 units (20 coupes and 20 roadsters).

- The Centenario features a naturally aspirated V12 engine.

- It showcases advanced aerodynamics and lightweight materials.

- Accelerates from 0 to 60 mph in under 2.8 seconds.

FERRARI 488 PISTA

- The 488 Pista is a track-focused variant of the Ferrari 488.

- Pista means "track" in Italian, highlighting its racing inspiration.

- It features a twin-turbocharged V8 engine with over 700 horsepower.

- The 488 Pista has extensive weight reduction measures for agility.

- Ferrari's Side-Slip Angle Control system enhances handling dynamics.

PORSCHE CARRERA GT

- **The Carrera GT is a mid-engine sports car produced by Porsche.**

- **It features a naturally aspirated V10 engine with a distinctive exhaust note.**

- **The Carrera GT's design was inspired by Porsche's racing heritage.**

- **Known for its carbon-fiber monocoque chassis for lightness and strength.**

- **Limited production, with around 1,270 units produced from 2004 to 2007.**

MCLAREN 600LT

- **The 600LT is part of McLaren's Longtail lineup.**

- **LT stands for "Longtail," emphasizing performance and aerodynamics.**

- **It features a twin-turbocharged V8 engine with 600 horsepower.**

- **The 600LT has lightweight carbon fiber components for agility.**

- **Known for its distinctive top-exit exhaust pipes.**

AUDI RS E-TRON GT

- **The RS e-tron GT is Audi's high-performance electric sports sedan.**

- **It shares its platform with the Porsche Taycan, emphasizing performance.**

- **Dual electric motors provide all-wheel drive and rapid acceleration.**

- **The RS e-tron GT features Audi's signature Quattro system.**

- **It combines sporty design with sustainable electric performance.**

PAGANI IMOLA

- **Imola is named after the famous Italian racing circuit.**

- **Limited to only 5 units worldwide, making it extremely exclusive.**

- **The Imola features extensive aerodynamic enhancements.**

- **Pagani emphasizes personalized design options for Imola buyers.**

- **It incorporates advanced materials for performance and safety.**

KOENIGSEGG REGERA

- **The Regera is a hybrid hypercar featuring a unique Direct Drive system.**

- **It combines a V8 engine with electric motors, producing over 1,500 horsepower.**

- **The Regera has a top speed exceeding 250 mph.**

- **It features a lightweight carbon fiber construction for performance.**

- **Limited production, with only 80 units planned.**

BUGATTI DIVO

- **The Divo is a track-focused variant based on the Bugatti Chiron.**

- **Named after Albert Divo, a French racing driver who won for Bugatti in the 1920s.**

- **It features aerodynamic enhancements for increased downforce.**

- **The Divo has a top speed lower than the Chiron but excels in handling.**

- **Limited production, with only 40 units produced.**

ASTON MARTIN DBS SUPERLEGGERA

- **The DBS Superleggera is a grand tourer with a focus on performance.**

- **Superleggera means "super light" in Italian, highlighting its lightweight construction.**

- **It features a twin-turbocharged V12 engine with over 700 horsepower.**

- **The DBS Superleggera has a sleek design and aggressive stance.**

- **Combines luxurious features with high-performance capabilities.**

LAMBORGHINI SIAN

- **The Sian is Lamborghini's first hybrid production car.**

- **Features a V12 engine combined with a supercapacitor-based hybrid system.**

- **The Sian's design is inspired by the iconic Lamborghini Countach.**

- **Limited to 63 units, with each having unique customization options.**

- **It has a top speed exceeding 220 mph.**

FERRARI F8 TRIBUTO

- **The F8 Tributo is a mid-engine sports car replacing the 488 GTB.**

- **Tributo pays homage to Ferrari's V8 heritage.**

- **It features a twin-turbocharged V8 engine with over 700 horsepower.**

- **The F8 Tributo has advanced aerodynamics for improved performance.**

- **Accelerates from 0 to 60 mph in under 2.9 seconds.**

PORSCHE 911 GT3 RS

- **The GT3 RS is a high-performance variant of the Porsche 911.**

- **It features a naturally aspirated flat-six engine with over 500 horsepower.**

- **The GT3 RS emphasizes track performance with aerodynamic enhancements.**

- **Known for its lightweight design and rear-wheel-drive configuration.**

- **The RS model undergoes extensive testing on the Nürburgring.**

MCLAREN GT

- **The GT is McLaren's grand tourer, combining comfort and performance.**

- **It features a twin-turbocharged V8 engine with over 600 horsepower.**

- **The GT has a lightweight carbon fiber structure for agility.**

- **McLaren prioritizes long-distance comfort with luxurious interiors.**

- **It offers a practical and spacious luggage compartment.**

RIMAC NEVERA

- **The Nevera is an all-electric hypercar from Rimac.**

- **It features four electric motors, each powering a wheel independently.**

- **Accelerates from 0 to 60 mph in under 1.85 seconds.**

- **The Nevera has a top speed exceeding 250 mph.**

- **Rimac focuses on advanced electric vehicle technology.**

LOTUS ELISE CUP 260

- **The Elise Cup 260 is a lightweight and track-focused version of the Elise.**

- **It features a supercharged four-cylinder engine with 250 horsepower.**

- **Known for its minimalist design and agile handling.**

- **The Cup 260 has aerodynamic enhancements for improved performance.**

- **Lotus emphasizes the Elise's connection to its racing roots.**

FORD GT MK II

- The GT Mk II is a limited-edition, track-only variant of the Ford GT.

- It features a 3.5-liter EcoBoost V6 engine with over 700 horsepower.

- The Mk II has enhanced aerodynamics for increased downforce.

- Limited production, with only 45 units produced.

- Designed for customers who want a high-performance track experience.

LAMBORGHINI VENENO

- **The Veneno is an extreme limited-production hypercar from Lamborghini.**

- **It features a naturally aspirated V12 engine with over 700 horsepower.**

- **Lamborghini produced only three coupe versions and nine roadsters.**

- **The Veneno's design is inspired by aerospace and racing aesthetics.**

- **Known for its distinctive and aggressive styling.**

BUGATTI CENTODIECI

- Centodieci means "one hundred ten" in Italian, celebrating Bugatti's 110th anniversary.

- It pays homage to the iconic Bugatti EB110 from the 1990s.

- The Centodieci features a quad-turbocharged W16 engine.

- It has a top speed electronically limited to 236 mph.

- Limited production, with only 10 units produced.

KOENIGSEGG AGERA RS

- **The Agera RS is known for its high top speed and quick acceleration.**

- **It set a record as the world's fastest production car, reaching 277.9 mph.**

- **Features a twin-turbocharged V8 engine with up to 1,341 horsepower.**

- **The Agera RS has advanced aerodynamics, including a dynamic rear wing.**

- **Koenigsegg focuses on lightweight construction and performance.**

PAGANI HUAYRA ROADSTER

- The Huayra Roadster is the open-top version of the Pagani Huayra.

- It features a twin-turbocharged V12 engine sourced from Mercedes-AMG.

- Each Huayra Roadster is custom-built to the buyer's specifications.

- The design incorporates advanced aerodynamics and lightweight materials.
- Limited production, with a small number of units produced.

ASTON MARTIN VULCAN

- **The Vulcan is a track-only hypercar from Aston Martin.**

- **It features a naturally aspirated V12 engine with over 800 horsepower.**

- **Aston Martin produced only 24 units of the Vulcan.**

- **The car offers a customizable driver development program.**

- **Known for its aggressive styling and high-performance capabilities.**

MCLAREN SABRE

- **The Sabre is a limited-production hypercar from McLaren.**

- **It features a twin-turbocharged V8 engine with over 800 horsepower.**

- **McLaren produced only 15 units of the Sabre, exclusively for the U.S. market.**

- **The Sabre has a unique design, emphasizing aerodynamics and exclusivity.**

- **It represents McLaren's commitment to extreme performance.**

RANDOM SUPERCAR FACTS

- The McLaren F1, introduced in the 1990s, held the title of the world's fastest production car for over a decade.

- The Ferrari Enzo was named after the company's founder, Enzo Ferrari, and only 399 units were produced.

- The Mercedes-Benz CLK GTR is a road-legal version of a car developed for GT1 class racing and is extremely rare.

- The Lamborghini Diablo was one of the first production cars to have a top speed exceeding 200 mph.

- The Maserati MC12 shares its platform with the Ferrari Enzo but has a longer wheelbase and unique bodywork.

RANDOM SUPERCAR FACTS

- The Porsche 959, produced in the 1980s, was considered one of the most technologically advanced cars of its time.

- The Bugatti EB110, introduced in the early '90s, was the first production car to use a carbon fiber monocoque.

- The Audi R8's design was inspired by the Audi Le Mans Quattro concept car, a prototype racer.

- The Pagani Zonda Cinque Roadster is known for its extensive use of carbon fiber and a sequential gearbox.

- The McLaren 675LT "Longtail" is named in honor of the McLaren F1 GTR Longtail race car.

RANDOM SUPERCAR FACTS

- The Aston Martin DB11 features an aerodynamic innovation called the "Curlicue" to reduce lift.

- The Rimac C_Two's electric powertrain can recover energy during braking with regenerative braking systems.

- The Koenigsegg Jesko Absolut variant aims for a top speed record, prioritizing straight-line speed.

- The Ford GT's flying buttresses not only contribute to aerodynamics but also house functional air intakes.

- The Lexus LFA's engine has a unique design, including titanium valves and a forged crankshaft.

RANDOM SUPERCAR FACTS

- **The Ferrari LaFerrari features an active aerodynamics system called "Aero Bridge" for improved downforce.**

- **The McLaren P1's interior incorporates lightweight materials, including a carbon fiber monocoque.**

- **The Lamborghini Aventador uses a pushrod suspension system inspired by Formula 1 cars.**

- **The Bugatti Chiron's quad-turbocharged W16 engine has 1,600 Nm of torque at 2,000 to 6,000 rpm.**

- **The Porsche 918 Spyder has a "Weissach Package" option, reducing weight with carbon fiber components.**

RANDOM SUPERCAR FACTS

- The McLaren Sabre has a bespoke aerodynamic package, individually tailored for each of its 15 owners.

- The Bugatti Centodieci's engine is an 8.0-liter quad-turbocharged W16, delivering 1,600 horsepower.

- The Koenigsegg Gemera is a "mega-GT" with four seats and a 1,700-horsepower hybrid powertrain.

- The Pagani Imola is named after the Imola circuit and limited to only 5 units, all of which were pre-sold.

- The Aston Martin Valkyrie's V12 engine was developed by Cosworth and can rev up to 11,100 rpm.

RANDOM SUPERCAR FACTS

- The Ferrari SF90 Stradale is the first Ferrari production car to feature all-wheel drive.

- The Rimac Concept_One features a complex torque vectoring system for precise control.

- The Lotus Evija is the first all-electric hypercar from Lotus, producing over 1,900 horsepower.

- The Audi RS e-tron GT is the first all-electric RS model from Audi, combining performance and sustainability.

- The Pagani Huayra Roadster uses a lightweight and strong material called Carbotitanium for its chassis.

RANDOM SUPERCAR FACTS

- The McLaren 765LT features a distinctive "Longtail" active rear wing for enhanced aerodynamics.

- The Porsche Carrera GT has a manually operated convertible top, emphasizing a purist driving experience.

- The Ford Mustang Shelby GT500's engine is hand-built and features a 2.65-liter supercharger.

- The Lamborghini Sian is the first hybrid Lamborghini, utilizing a supercapacitor for energy storage.

- The Rimac Nevera's electric motors generate a combined 1,914 horsepower and 2,360 Nm of torque.

RANDOM SUPERCAR FACTS

- The Lotus Elise Cup 260 has a power-to-weight ratio of 238 horsepower per ton.

- The Lamborghini Veneno's hexagonal theme in its design is inspired by the molecular structure of carbon.

- The Bugatti Divo can generate 1.6 lateral g-force, making it exceptionally agile in corners.

- The Porsche 911 GT3 RS uses a magnesium roof to reduce weight and lower the center of gravity.

- The Ford GT Mk II's aerodynamics are optimized for maximum downforce, featuring a large rear wing.

RANDOM SUPERCAR FACTS

- The Aston Martin One-77's chassis is made from a carbon-fiber monocoque, contributing to its lightweight construction.

- The Rimac C_Two features facial recognition technology to unlock the car, enhancing security.

- The Ford Mustang Shelby GT500 has a dual-clutch transmission, a first for a Shelby Mustang.

- The McLaren 600LT's exhaust pipes exit through the top of the rear deck, creating a distinct visual feature.

- The Lamborghini Huracan Performante set a Nürburgring lap record for production cars in 2017.

RANDOM SUPERCAR FACTS

- The Pagani Zonda's name pays homage to a wind over the Andes mountains, emphasizing aerodynamics.

- The Koenigsegg Jesko's lightweight carbon fiber wheels contribute to its overall performance and agility.

- The Bugatti Chiron's quad-turbo W16 engine inhales 60,000 liters of air per minute at full throttle.

- The Porsche 911 GT2 RS holds the record for the fastest production car lap time at the Nürburgring Nordschleife.

- The Ferrari 488 Pista's aerodynamics are inspired by the 488 GTE and 488 Challenge racing cars.

Printed in Great Britain
by Amazon